AF421216

LABORATORY MANUAL ON
SOIL PHYSICAL PROPERTIES

LABORATORY MANUAL ON SOIL PHYSICAL PROPERTIES

M. Singa Rao

Professor of Soil Physics, Department of Soil Science & Agricultural Chemistry

Prabhu Prasadini

Associate Professor & Head, Department of Environmental Science & Technology

and

M. Jeevaratna Raju

Assistant Professor, Department of Soil Science & Agricultural Chemistry

Acharya N.G. Ranga Agricultural University

College of Agriculture, Rajendranagar

Hyderabad, Andhra Pradesh

BSP **BS Publications**

4-4-309, Giriraj Lane, Sultan Bazar,
Hyderabad - 500 095 A.P.
Phone : 040 - 23445601, 23445688
e-mail : contactus@bspublications.net

Published by :

BSP **BS Publications**

4-4-309, Giriraj Lane, Sultan Bazar,
Hyderabad - 500 095 A.P.
Phone : 040 - 23445601, 23445688
Fax : 040 - 23445611
e-mail : contactus@bspublications.net

Printed at

Adithya Art Printers
Ph : 23445605

Price : Rs. 70.00
ISBN : 81-7800-094-6

CONTENTS

1. Concept of Analysis 1

2. State Properties of Soils 7

    2.1 Particle Size Analysis 9

    2.2 Particle Density 21

    2.3 Bulk Density 25

    2.4 Determination of Physical Constants of Soil 31

    2.5 Aggregate Analysis 35

    2.6 Penetrability 41

    2.7 Soil Color 47

3. Soil Hydraulic Properties 51

    3.1 Soil Water Content by Gravimetric Method 53

    3.2 Soil Water Content Using Neutron Probe 57

    3.3 Soil Water Potential 61

    3.4 Soil Moisture Constants 67

    3.5 Water Intake Rate 73

    3.6 Saturated Hydraulic Conductivity 79

4. Thermal Properties 85

    4.1 Soil Temperature 87

Chapter 1

Concept of Analysis

1

Concept of Analysis

Exercise - 1

Analysis is detailed examination of the elements or structure of something, typically as a basis for discussion or interpretation. The first step in any analysis is the identification of soil properties to be determined relevant to the situation/problem. These may by physical, chemical, biological or their combination. The next step is determination/estimation of the property. Selection of representative site and collection of sample is very important.

While analyzing a soil property basic knowledge on dimensions and units is necessary. Some commonly used dimensions and units are given in Table 1.1.

Table 1.1 Dimensions and Units

Quantity	Dimension	Units	
		CGS	SI
Length	L	cm	m
Mass	M	g	Kg
Time	T	s	s
Temperature	^{o}	^{o}c	^{o}K
Area	L^2	cm^2	m^2
Volume	L^3	cm^3	m^3
Density	M/L^3	g/cm^3	Mg/m^3 *
Force	ML/T^2	$g\,cm/s^2$, dyne	$Kg\,m/s^2$, N*
Pressure	M/LT^2	dynes/ cm^2, bars	N/m^2, Pa*
Surface tension	M/T^2	g/s^2, dynes/ cm	Kg/s^2, N/m
Energy	$M\,L^2/T^2$	$g\,cm^2/s^2$, ergs	$Kg\,m^2/s^2$
Potential	L^2/T^2	cm^2/s^2, ergs/g	J/Kg*
Diffusivity	L^2/T	cm^2/s	m^2/s
Viscosity	M/LT	g/cm-s, poise	Kg/m-s
Coductivity	L/T	cm/s	m/s
Permeability	L^2	cm	

* Mg = Mega gram;   N = Newton;   Pa = Pascal;   J = Joule.

In this manual analysis of some of the soil physical properties is covered. Knowledge of soil physical properties is a basic necessity in soil, water and nutrient management availability and movement of nutrients and water to plants is fundamentally decided by soil physical condition.

Chapter 2

State Properties of Soils

Exercise - 2.1

Particle Size Analysis

Sand, silt and clay are the primary particles comprising the soil solids. These are generally clustered together as secondary particles. The grouping of particles based on the size are given in   Table 2.1. The procedure of separating out these primary particles and measuring their proportions is called particle size analysis or mechanical analysis. The relative proportion of size groups of sand, silt and clay in a soil is called soil texture. Soil texture helps in understanding all phenomena of soils such as infiltration, hydraulic conductivity, water retention, plasticity, cation exchange capacity, workability *etc*. The texture of soil does not change by normal management practices.

Table 2.1 The International Society of Soil Science (ISSS) and USDA has grouped soil particles into the following classes based on size

System Fraction	ISSS Diameter, mm	USDA Diameter, mm
Very coarse sand	-	2 - 1
Coarse sand	2 - 0.2	1 - 0.5
Medium sand	-	0.5 - 0.25
Fine sand	0.2 - 0.02	0.25 - 0.1
Very fine sand	-	0.1 - 0.05
Silt	0.02 - 0.002	0.05 - 0.002
Clay	<0.002	<0.002

Aim :

Determination of relative proportions of sand, silt and clay in a given soil.

Principle :

Two important steps are involved in particle size analysis are, (i) dispersion of soil solids and (ii) their fractionation. Stoke's law is the basic principle involved in the fractionation of finer particles. Two methods commonly followed in particle size analysis are - International pipette method or Robinson pipette method, and Bouyoucos hydrometer method or hydrometer method. In the pipette method weight of the particles in known amount of soil suspension whereas density of the suspension is determined, is measured in the hydrometer method. Hydrometer method is simple and rapid whereas pipette method is more accurate but is laborious and time consuming.

Stoke's law (1851) :

The resistance offered by the liquid to a freely falling insoluble spherical smooth particle varies with square of the radius of the particle but not with the surface area of the particle.

Stoke's equation for settling velocity :

A rigid smooth spherical particle falling freely through a liquid of a lower density will attain a constant velocity, called terminal velocity when the downward and upward forces acting on the particle are equal.

$$F_{downward} = \text{(force of gravity)} - \text{(bouyant force offered by the liquid)}$$

$$= (4/3 \; \tilde{O} \, r^3 \, r_s \, g) - (4/3 \; \tilde{O} \, r^3 \, r_l \, g)$$

$$= 4/3 \; \tilde{O} \, r^3 . \, (r_s - r_l). \, g \qquad \qquad \dots\dots(2.1)$$

Where r is radius of the particle, r_s is density of particles, r_l is density of liquid and g is acceleration due to gravity

$$F_{upward} = \text{resistance to fall offered by the liquid to the sphere} = \text{viscous force}$$

$$= 2 \; \tilde{O} r \, hn \times 3$$

$$= 6 \; \tilde{O} r \, hn \qquad \qquad \dots\dots(2.2)$$

Where h is the co-efficient of viscosity of the liquid and n is velocity of particle.

When the magnitude of these two forces are equal and act in opposite directions, particle moves with constant velocity

$$F_{downward} = F_{upward} \qquad \qquad \dots\dots(2.3)$$

After rearranging, we get settling velocity as

$$n = \frac{2}{9} \; \frac{gr^2(\rho_s - \rho_l)}{\eta} = \frac{h}{t} \qquad \qquad \dots\dots(2.4)$$

$$n = K \, r^2 \text{ (where K is constant)}$$

$$n \text{ a } r^2$$

where h is depth of settling. The equation (2.4) is called the Stoke's equation

Thus, in a time t all particles having settling velocities greater than h/t would have travelled beyond depth h and the particles having settling velocities lesser than h/t would be yet present above this depth i.e., within h cm. The settling times of silt and clay at different temperatures are shown in Table 2.2, for example at 30 °C all particles coarser than 20m takes 3 minutes 50 seconds to travel a depth of 10 cm.

Procedure :

There are three steps – A. Dispersion, B. Fractionation and C. Texture identification

A Dispersion

Primary soil particles are grouped and cemented together by various cementing materials like organic matter, calcium carbonate *etc.* present in soils. To determine relative distribution of soil particles ultimate soil particles, disperse the soil particles completely and is accomplished as described below :

Removal of cementing agents

Organic matter : It is removed by treating the soil with hydrogen peroxide which oxidizes the organic matter to carbon dioxide and water.

Oxides of iron and aluminum : It is different to eliminate the cementation influence of the oxides of iron and aluminum. The oxides are inactivated by reduction with sodium dithionate in sodium citrate which convert them into soluble forms.

Clay and clay cementation : Clay particles results in strong bonding upon dehydration. Thus these bonds can be broken by rehydration. The physical methods of dispersion after rehydration are shaking and stirring with an electric shaker.

Calcium carbonate : Cementation due to calcium carbonate is removed by treating the soil sample with dilute HCl.

Electrocytes : To deflocculate the soil particles, the soil is repeatedly washed with distilled water till it is free from soluble salts.

The HCl treatment saturates the exchange complex with hydrogen ions (H^+) which also keeps the soil flocculated. To disperse the soil, the H^+ in the soil exchange complex is replaced by a peptizing cation like Na^+ by treating the sample with sodium hexa meta phosphate $\{(Na\ PO_3)\}_6$ and dispersed by vigorous stirring of the soil water suspension with the electric shaker.

B   Fractionation

Two methods are generally followed – Pipette method and Hydrometer method.

PIPETTE METHOD

Coarse sand particles are separated by sieving the dispersed soil suspension. Fine sand particles are separated by decantation. Fractionation of the finer particles (clay and silt) is done based on the differences in their settling velocities in a liquid which is governed by Stoke's law of sedimentation.

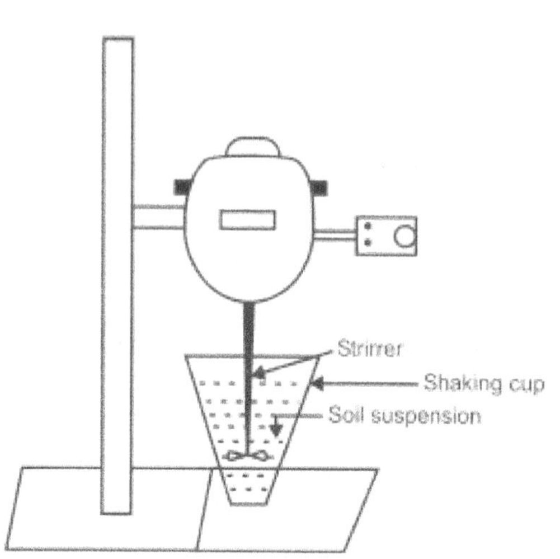

Fig. 2.1 Electric strirrer

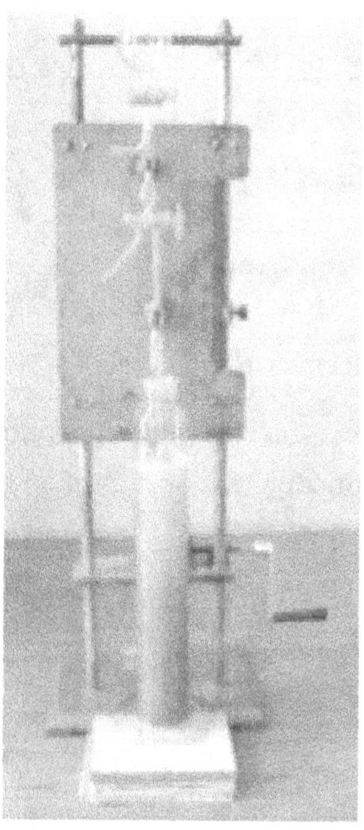

Fig. 2.2 Robinson pipettee.

Apparatus :

Tall form glass beaker of 600 ml capacity, hot water bath or hot plate, a glass rod with rubber sleeve on one end, wash bottle, Whatman No. 50 filter paper, electric stirrer   (Fig. 2.1) 2mm-mesh sieve, sedimentation cylinder of one liter capacity, glass funnel,   thermometer, plunger – an iron rod attached to a circular metallic disc of 5-6 cm in diameter and 0.16 cm in thickness, sampling pipette, beakers (100ml capacity), balance, oven, Robinson pipette (Fig. 2.2).

Reagents : H_2O_2 (Hydrogen peroxide) 30%, HCl (2N), $AgNO_3$ (N/10), NaOH (N/10), phenolphthalein indicator and sodium hexa meta phosphate(5% solution).

Procedure :

- Take 50 g air dry soil (ground and sieved through a 2-mm sieve) in a 600 ml tall form beaker and add about 200 ml distilled water.

Dispersion

- Add 4-5ml of 30 % H_2O_2 and swirl the contents well.

- Allow the reaction to take place for 5-10 minutes.

- Heat the contents an a hot water bath or on a hot plate while stirring the contents all the time with a glass rod to minimize frothing, till completion of the reaction. If frothing and reaction persists for a long period another lot of H_2O_2 should be added.

- Cool the contents wash down the soil particles from the inner sides of the beaker by rubbing with the glass rod and using a jet of distilled water.

- Add 30-40ml of 0.1N HCl. If more than 2% $CaCO_3$ is present, more HCl should be added at the rate of 2.5 ml for each per cent of $CaCO_3$. This step can be omitted if soil is free from $CaCO_3$.

- Leave it for about an hour with intermittent shaking.

- Filter the contents through Whatman No. 50 filter paper.

- Discard the filtrate.

- Wash the soil retained on the filter paper with distilled water till the filtrate is free from chlorides. (This can be tested by adding a small amount of $Ag NO_3$ to the filtrate. If a white precipitate appears, chloride is still present and sample needs more washing).

- Transfer back the soil from filter paper to a 600 ml beaker with a jet of distilled water.

- The volume is to be madeup to about 300 ml with distilled water.

- Then few drops (5-6) of phenolphthalein indicator is to be added.

- Add N/10 NaOH till the whole suspension shows a pink colour indicating its alkaline reaction. Avoid over-alkalization.

- Add 10ml of 5% sodium hexa meta phosphate, stir the contents and leave it for half an hour.

- Shake the contents of the beaker with an electric stirrer for 10 minutes.

Fractionation

- Transfer the contents to one liter cylinder.

- Make the volume of the suspension in the cylinder to one litre by adding more distilled water.

- Note down the temperature of the suspension.

Table 2.2 Sedimentation times for silt and clay particles in water to reach at 10 cm depth (for ISSS system) at different temperatures of solution

Temperature	Settling time			
	Clay (<2 m)		Silt + Clay (20m)	
^{0}C	Hours	Minutes	Minutes	Seconds
15	9	5	5	30
16	8	50	5	20
17	8	35	5	10
18	8	25	5	0
19	8	10	4	50
20	8	0	4	45
21	7	50	4	40
22	7	35	4	35
23	7	25	4	30
24	7	15	4	20
25	7	0	4	15
26	6	55	4	10
27	6	45	4	5
28	6	40	4	0
29	6	30	3	55
30	6	20	3	50
31	6	15	3	45
32	6	0	3	40

- Against this temperature, read the required time for sampling of (silt+ caly) and clay alone from Table 2.2.

- Stir the contents with plunger by moving it up and down gently for about 20-25 times in one minute so that no soil remains settled at the bottom of the cylinder. This is the beginning time for settling of particles.

- Insert the sampling pipette gently into the suspension and dip it to 10 cm depth from the surface of suspension about 20 seconds before the expiry of the sampling time.

- Pipette out 25ml at each requisite time (for silt + clay and clay) at a moderate speed.

- Take out the pipette and transfer the sample collected to a pre-weighed 100 ml beaker/ crucible.

- Oven dry the sample at 105 ^{o}C to a constant weight.

- Weigh the beakers with (silt + clay) and clay up to milligrams. Note down these weights.

- Calculate the weight of silt by subtracting weight of clay from that of (silt + clay).

- Calculate the sand content by subtracting weight of (silt + clay) from the total sample weight.

Observations and Calculations :

Weight of soil sample taken = W g

Weight of empty beaker = w_1 g

Weight of beaker + (silt +clay) = w_2 g

Weight of beaker + clay = w_3 g

$$\text{Percent clay} = \left(\frac{w_3 - w_1}{W}\right) \times 100$$

$$\text{Percent (silt + clay)} = \left(\frac{w_2 - w_1}{W}\right) \times 100$$

Percent silt = % (silt + clay) – (% clay)

Percent sand = 100 – % (silt + clay)

Precautions :

- H_2O_2 is oxidizing agent and should be handled carefully.
- Do not add excess alkali than required, otherwise it will initiate flocculation instead of dispersion.
- Temperature fluctuations should be avoided.

HYDROMETER METHOD

Apparatus : Hydrometer (Fig. 2.3) and others mentioned as under pipette method.

Hydrometer meant for particle size analysis was introduced by Bouyoucos in 1927. This is made up of glass having bulb at its bottom and lengthy stem on its top. The total length of this hydrometer is 28.5 cm and the length of its stem is 15 cm. A dead weight, either mercury or lead, is added at the bottom of the bulb to make the hydrometer to float vertically in the suspension. The scale drawn on the stem at 68°F or 19.4°C is calibrated in grams per liter to give densities of the liquids.

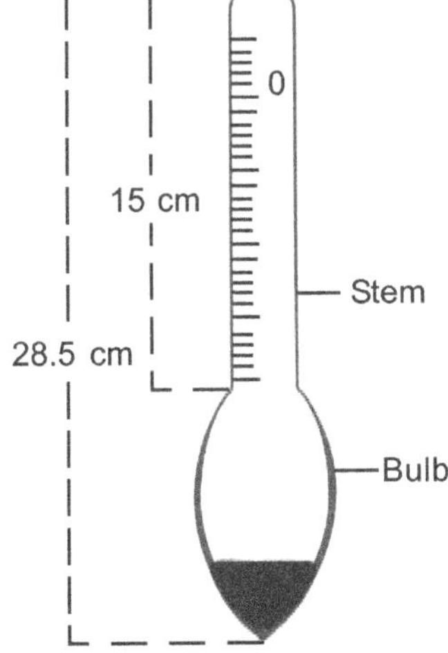

Fig. 2.3 Bouyoucos Hydrometer.

Procedure:

- Take 50 g air dry soil (2mm sieved)
- Disperse the sample (as under pipette method)
- Transfer the dispersed soil to the sedimentation cylinder and makeup the volume to one liter with distilled water.
- Shake the suspension top to bottom or with a plunger about 20 times a minute.
- Take hydrometer readings at 40 seconds & 2 hours in USDA system and 4 minutes & 2 hours in ISSS system by lowering the hydrometer gently into the suspension about 20 seconds, before each reading for (silt + clay) and clay separately.
- Note down the temperature of the suspension and apply temperature correction to the hydrometer readings.
- For taking blank reading follow the same procedure except that the soil sample is not there.

Note : Temperature correction to be made

Bouyoucos (1927) considered the hydrometer readings taken at 4 min and 2 hours to calculate (silt+clay) and clay content, respectively, in a soil sample. He recommended the temperature to be maintained at 68^0 F (19.4 $^{\circ}$C). If temperature varies, add 0.2 to the hydrometer reading (HMR) for each degree rise above 68°F and subtract 0.2 for each degree fall below 68°F within 60 to 75°F. (Add 0.3 to the HMR for each 1 $^{\circ}$C rise and subtract 0.3 for each 1 $^{\circ}$C fall in temperature if HM scale is calibrated at 19.4 $^{\circ}$C).

Table 2.3 Observations and calculations for ISSS system

	Hydrometer reading (HMR)	Temperature	Temperature correction	Corrected HMR (CHMR)
Blank				
At 4 minutes				
At 2 hrs				

$$\% \ (silt + clay) = \frac{CHMR \ after \ 4 \ min - Blank}{Weight \ of \ soil} \times 100$$

$$\% \ clay = \frac{CHMR \ after \ 2 \ hrs - Blank}{Weight \ of \ soil} \times 100$$

$$\% \ silt = \% \ (silt + clay) - \% \ clay$$

$$\% \ sand = 100 - \% \ (silt + clay)$$

C Identification of soil texture

The texture of soil is identified from the relative proportions of sand, silt and clay. Two systems of soil texture classifications, as suggested by USDA and ISSS Figure), are in common use. Both make use of equilateral triangle whose area is divided into 12 compartments, each representing a textural class. The difference between the two is primarily due to difference in size ranges of sand and silt fraction. For the determination of the texture of a soil, locate the clay and the silt percentages on the respective sides of the triangle. Draw a line inward parallel to the sand side in the former case and parallel to the clay side in the latter case. The compartment in which the two lines intersect is the texture of the soil.

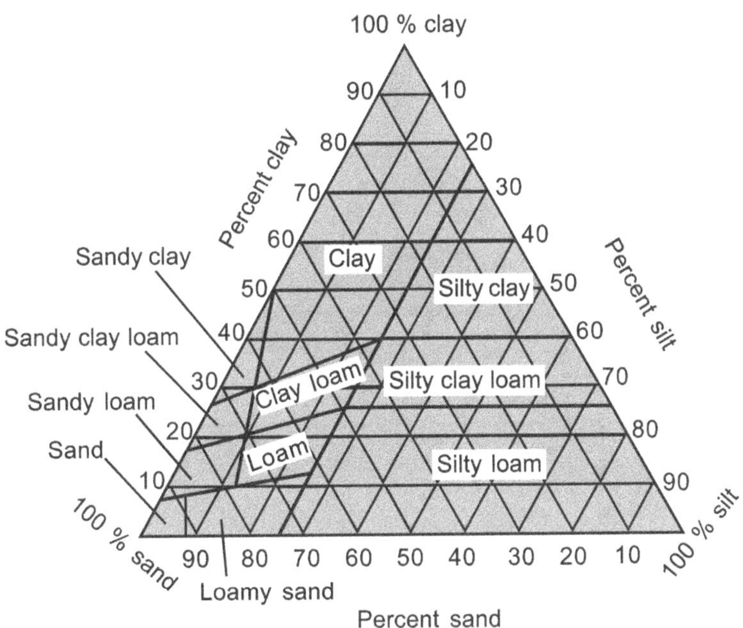

Fig. 2.4 Textural Diagram (ISSS)

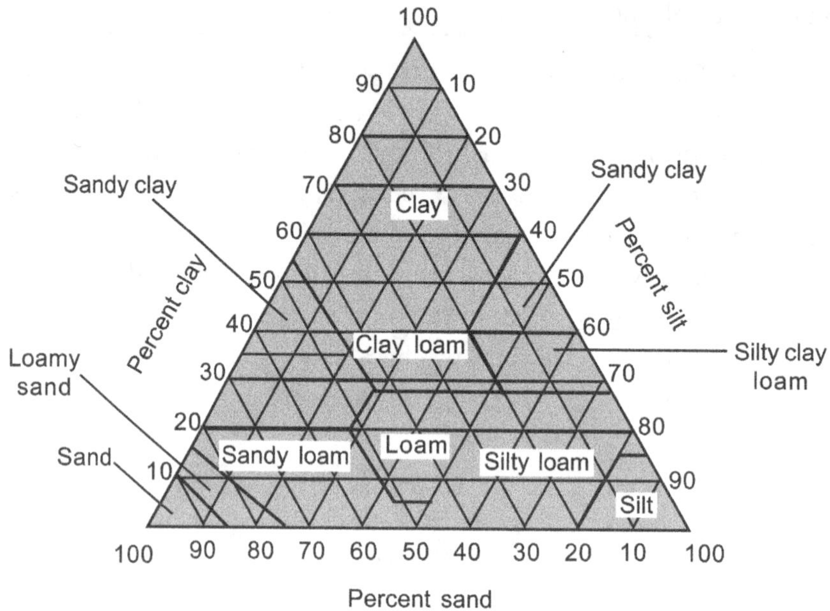

Fig. 2.5 Textural diagram (USDA).

Result :

	Percent	
	Pipette method	**Hydrometer method**
Sand		
Silt		
Clay		
Texture		

Exercise - 2.2

Particle Density

The particle density of a soil is defined as the ratio of mass of soil particles and their volume.   It is expressed as g cm^{-3} which is equivalent to Megagrams per cubic metre (Mg m^{-3}) in SI units. In most of the mineral soils it varies between 2.60 and 2.70 Mg m^{-3} with an average value of 2.65 Mg m^{-3}. The presence of large amount of organic matter may decrease and that of heavy minerals like hornblende and zircon may increase the particle density value. A knowledge of particle density is of great significance in the volume-mass relationships of the soil *e.g.* bulk density, porosity and void ratio. Further, the particle density is used in calculating settling velocities of soil separates of different sizes in particle size analysis. It also gives idea about soil erosion.

Aim :

Determination of particle density using pycnometer.

Principle :

To find out the volume of soil particles of known mass of dry soil it is immersed in water. The amount of displaced water is equal to the volume of soil particles, *i.e.* solid phase of the soil sample.

Apparatus :

pycnometer (Fig. 2.6), 20-ml pipette, balance, hot plate, filter paper.

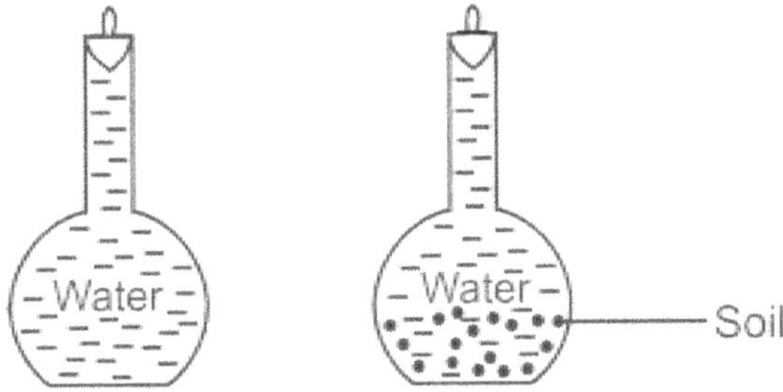

Fig. 2.6 Pycnometer.

Procedure :

- Take a clean empty and dry specific gravity bottle along with its stopper (pycnometer) and weigh it.
- Introduce approximately 10g of oven dried 2mm sieved soil in the pycnometer and note down the weight along with the stopper.
- Fill the pycnometer to about half with water and wash soil particles sticking to the inner side of the neck with a jet of water.
- Expel the entrapped air by shaking and gentle boiling of the contents.
- Allow the contents to cool to room temperature and fill the pycnometer to the brim with boiled and cooled distilled water.
- Fix the stopper well.
- Clean the outer side of the pycnometer with a filter paper and weigh it.
- Remove the contents of the pycnometer *i.e.*, soil and water and clean it with distilled water and fill it with boiled and cooled distilled water and replace the stopper.
- Wipe outer side of the pycnometer with a piece of filter paper and weigh it.

Precautions :

- Soils rich in organic matter should be treated first with H_2O_2 to oxidize organic matter.
- Use deaerated distilled water.

Observations and Calculations

Weight of empty pycnometer along with stopper, g = w_p

Weight of the pycnometer along with stopper + soil, g = w_{ps}

Weight of pycnometer + soil + water, g = w_{psw}

Weight of water-filed pycnometer, g = w_{pw}

Volume of soil solids, V_s in cm^3 = volume of water equal to the volume of soil solids

$$= (w_{pw} - w_p) - (w_{psw} - w_{ps}) = w_{pw} - w_p - w_{psw} + w_{ps}$$

$$\text{Particle density, g cm}^{-3} \text{ or Mg m}^{-3} = \frac{\text{Mass of the oven dried sample}}{\text{Volume of the solids excluding pore space}}$$

$$= (w_{ps} - w_p) / (w_{pw} - w_p - w_{psw} + w_{ps})$$

Result :

Particle density of the given soil is ....

Exercise - 2.3

Bulk Density

Bulk density is an important soil parameter. It is defined as the ratio of mass of oven dry soil to its bulk volume. It is expressed as $g\ cm^{-3}$ or $Mg\ m^{-3}$. The bulk volume is the volume of soil particles plus pore space. Bulk density of soil depends upon the soil texture, structure, mineralogical composition, organic matter content, depth of soil and soil management practices like tillage. Bulk density values are used to convert water content on mass basis to volume basis, for calculating porosity and void ratio. It is a simple index for soil structure. In general, it ranges from 1.1 to 1.4 $g\ cm^{-3}$ in fine textured soils and 1.5 to 1.8 $g\ cm^{-3}$ in coarse textured soils. It is measured by different methods *viz.,* core method, weighing bottle method, clod saturation method, clod coating method, scoop method and densitometer method. Clod method usually gives higher bulk density values than do other methods.

Aim :

Determination of bulk density of soil.

CORE METHOD

Principle :

A metallic core of known volume is driven into the soil and an undisturbed soil sample is taken. The soil sample is weighed after oven drying the soil. The volume is calculated from the core dimensions used for drawing the sample. By substituting the values in the formula bulk density is calculated.

Apparatus :

core sampler (Fig. 2.7), hammer, oven, balance, aluminum boxes, and knife.

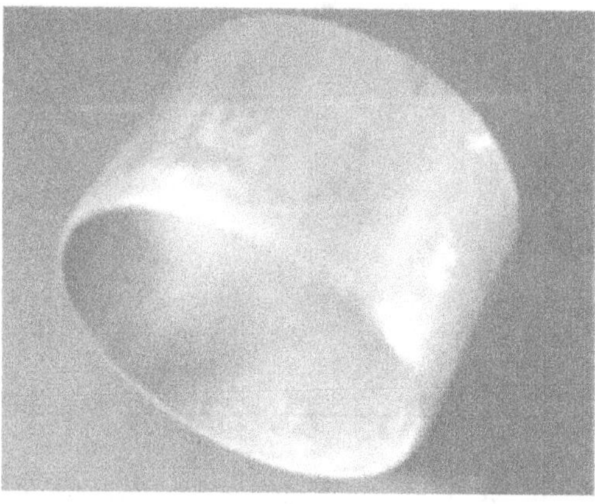

Fig. 2.7 Core for collection of sample.

Procedure :

- Smoothen the surface or remove the loose surface soil, where bulk density is to be determined.
- Drive the metallic core into the soil using a small hammer.
- Carefully remove the core containing soil sample by excavating soil form the sides, using a pick axe or crow bar.
- Trim the protruding soil with the knife on either side of the core.
- Transfer the sample from the core to the pre-weighed aluminum box.
- Dry the soil in oven at 105 ^{o}C for 24 hours, cool it and determine oven-dry weight of the sample.
- Measure inner dimensions i.e. radius and depth of the core and find its volume.

Precautions

- Always use dent-free cores.
- Core should be driven straight (non- inclined) into soil.
- Core should be driven gently to avoid shattering or compaction of soil.
- Sample should be dried at temperature not exceeding 105°C till constant weight is obtained.

Observations and Calculations :

Weight of the empty aluminum box, g = w_b

Weight of the aluminum box with oven dried soil, g = w_{bs}

Diameter of the core, cm = d

Radius of the core, cm = d/2 = r

Length / depth of the core, cm = h

Volume of the core, $cm^3 = V = P r^2 h$

Bulk density, $g / cm^3 = \dfrac{w_{bs} - w_b}{V}$

Results :

Bulk density of the soil is________

CLOD METHOD

Principle :

Clod method is used to find out the bulk density of clods or bigger aggregates.   The volume of the clod may be determined by coating the clod of known weight with water repellent substances like paraffin wax. The clod is weighed first in air and then again while immersed in a liquid of known density (water) .The volume is known by making use of Archimedes principle *i.e.* the loss in weight when weighed in water is equal to the weight of volume of the water displaced by the clod. The clod must be sufficiently stable to cohere during coating, weighing and handling.

Apparatus :

Balance - modified to accept the clod suspended below the balance arm by means of a nylon thread to allow weighing the clod when it is suspended in a container of liquid   (Fig. 2.8); paraffin wax; nylon thread; 500 ml beaker half filled with water, a small wooden bench to keep the beaker above the pan without touching the pan.

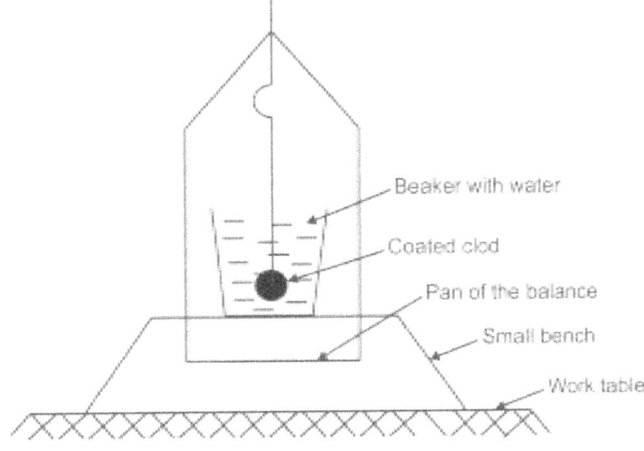

Fig. 2.8 Arrangements for weighing the coated clod in water.

Procedure :

·   Secure the clod with a long nylon thread, leaving sufficient thread to suspend clod to the balance arm.

·   Weigh the clod with the thread.

·   Melt the paraffin wax in a beaker.

·   Dip the clod for a while in the melted paraffin wax by holding the thread, then remove the clod and dry it. The clod should be water proof after paraffin coating.

·   Weigh the clod with its coating and the thread

·   Keep the wooden bench over the left pan of the balance without touching the pan.

·   Keep the 500 ml beaker half filled with water on the wooden bench.

·   Suspend the clod an the balance arm and weigh the clod when it is immersed in water. The clod should not touch the bottom or the sides of beaker.

·   Calculate the bulk density of clod.

·   To obtain a correction for water content of the soil, break open the clod, remove an aliquot of soil and determine the moisture content

Precautions :

·   Clod should be taken which represents the soil.

·   Insufficient or over coating of paraffin on the clod should be avoided.

Observations and calculations :

Weight of dry clod, g = W_{dc}

Weight of coated clod in air, g" = W_{cca}

Weight of coated clod in water, g = W_{ccw}

Loss in weight when weighted in water (Lw), g = $W_{cca} - W_{ccw}$

Volume of coated clod, V_{cc} = L_w / r_w

(r_w is density of water = 1.0 g cm^{-3})

Weight of coated material. g = $W_{cca} - W_{dc}$

Density of coated material, g cm^{-3} = r_{cm}

(Given on the paraffin wax tin)

Volume of coated material, cm^{-3} V_{cm} =

Volume of dry clod (V_{dc}), cm^{-3} = $V_{cc} - V_{cm}$

B.D, g cm^{-3} = $\dfrac{W\,dc}{V\,dc}$

Result : Bulk density is .... g cm^{-3}

Exercise - 2.4

Determination of Physical Constants of Soil

The physical constants of soil like bulk density, particle density, porosity and maximum water holding capacity (WHC) can be determined with the help of Keen cup. This method was developed by Keen Roezkowski in 1905. When all the pores of soil are filled with water the soil is said to be at its maximum WHC. The free energy of water at that stage is zero.

Aim :

To determine the physical constants of processed soil.

Apparatus :

Keen cup (Fig. 2.9), filter paper, glass rod/knife, balance, water trough, oven and Vernier callipars/ scale

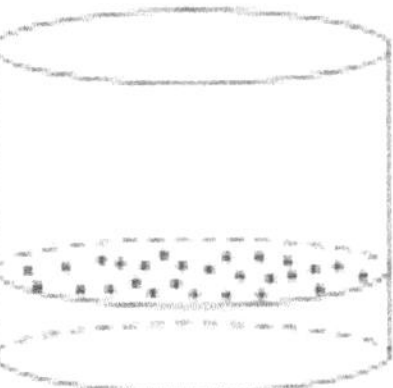

Fig. 2.9 Keen cup.

Principle :

Ratio of mass and bulk volume of soil sample including pore spaces gives bulk density whereas Ratio of mass and volume of soil solids gives particle density. Porosity of the soil can be determined from bulk density and particle density values. The soil being saturated gives information about maximum water holding capacity.

Procedure :

· Take a clean and dry Keen cup. Fix a filter paper which is cut to the size of the cup in circular shape in it at its bottom.

· Take the weight of dry Keen box along with filter paper in it.

· Take the soil sample passed through 2 mm sieve in the box giving small tappings.

· Fill the box till it is completely full.

- Remove the excess soil with a glass rod or spatula.
- Keep the Keen cup in a trough of water. Adjust the water level in such a way it is 2cm from the bottom of Keen cup in water. Allow it to saturate for 24 hours.
- Remove the excess amount of soil using a sharp knife/glass rod.
- Take out the box from the trough and allow the excess water to drain out for half an hour.
- Note down the weight of the Keen cup with wet soil.
- Keep the sample for drying along with the box in an oven at 105°C till a constant weight is obtained.
- Note down the weight of the box with dry soil.
- Note down the inner diameter and height of the Keen cup with Vernier callipars or scale.

Precautions :

- Wipe the outer side of the Keen box to remove soil and water before weighing it.
- Ensure even fixation of the mesh.
- Take care to maintain the water level below the brim and above the mesh bottom of the Keen cup, approximately 3/4th of total height of the cup, while saturating the soil.

Observations and Calculations :

Weight of the Keen cup + filter paper, g = w_b

Weight of the Keen box + filter paper + air dry soil, g = w_{bs}

Weight of the Keen box + filter paper + wet soil, g = w_{bws}

Weight of the Keen box + filter paper + oven dried soil, g = w_{bds}

Inner diameter of the Keen box, cm = d =

Inner radim of Keen box, cm = r =

Inner depth of the Keen box, cm = h =

Volume of the Keen box, cm^3 = V = pr^2 h

Bulk density =   g cm^3

Particle density =   g cm^3

Porosity =   × 100 or   × 100

Maximum water holding capacity =   × 100

Result :

Buk density is……, Particle density is……., % Porespace is ………,

Maximum water holding capacity ……….

Exercise - 2.5

Aggregate Analysis

The primary soil particles tend to group themselves into secondary units called soil aggregates. Soil aggregation is caused by the presence of cementing agents such as clay, organic matter, oxides of iron and aluminum and calcium carbonate. Aggregation is a dynamic property of soil and determines soil erosion, thermal regime and movement and retention of water and air aggregates disintegrate under the beating action of rain drops or irrigation and cause surface sealing and eventually crusting. The rate of disintegration depends upon the aggregate stability. In a soil, the amount of different-size aggregates and their stability are determined by wet- and dry-sieving methods. The former method is of particular importance in measuring water stable aggregates. Choice of the method depends on the objective of the analysis, where the objective of the work is to findout the resistance of soil to water erosion or surface crusting, wetting the soil by immersion method is preferred.

Aim :

To determine the aggregate size distribution using Yoder's apparatus.

Principle :

A known amount of the soil sample collected from the field is immersed in water for a short period of time for wetting. The wetted sample is sieved through a nest of sieves in Yoder's apparatus, which raises and lowers the nest of sieves. This is done to simulate the disruptive forces of water and facilitate the sieving of water stable aggregates through sieves. After about 30 minutes, sieves along with aggregates are removed and dried in an oven at $105^{\circ}C$. The dry aggregates are collected and weighed. The weight of these aggregates also includes weight of primary particles or sand fraction of respective sieve sizes. Therefore weight of the sand fraction is to be deducted from the mass of aggregates to get correct estimate of aggregates.

Apparatus :

Nests of sieves (5, 2, 1, 0.5, 0.25 and 0.1mm), each sieve 20 cm in diameter and 5 cm in height; 8 mm sieve; Yoder's apparatus (Fig. 2.10) which raises and lowers the nests of sieves in water through a height of 3.75 cm approximately 30 times per minute; Mechanical stirrer; Hydrometer (ASTM-1524 gram per litre type); Thermometer.

Fig. 2.10 Yoder's apparatus

Reagents :

Hydrogen peroxide and sodium hexa meta phosphate

Procedure :

- Collect the sample when the soil is moist and friable. Prepare the sample by gently breaking the clods so as to pass through 8 mm sieve but retained on 5 mm sieve.

- Air dry the sample collected over 5 mm sieve. Take three representative 50 g sub-samples. Oven dry one sample for moisture determination. Use the other two air dry samples (as two replicates) in following determinations.

- Distribute the air-dry sample evenly over the top (5mm) sieve.

- Wet the sample by spraying 10 cc of salt free water and leave it for 30 minutes for uniform distribution of moisture.

- Stack the nest of sieves in descending order of sive size and suspend these in the Yoder's tank. Raise the racks to maximum height.

- Fill the container with good quality water to a level slightly below that of the top sieve (5 mm size).

- Shake the sieves for 30 minutes. During shaking, aggregates of different sizes will be collected on different sieves.

- Remove the nests of sieves from the water tank keep them in an inclined position and put in an oven for drying at low temperature for an hour for the quick removal of the free water.

- Transfer the contents of sieves to pre-weighed aluminum boxes or glass beakers, dry them to a constant weight at 105^0C and determine the weight of aggregates.

- After getting the individual weights of different sizes collected on different sieves, mix them and follow the procedure of mechanical analysis for finding the sand fraction by using sieves or hydrometer after following the steps of dispersion as explained in exercise 2.1 (*i.e.,* addition of H_2O_2 and sodium hexa meta phosphate and stirring with mechanical stirrer).

Precautions :

- Wetting of the soil sample should be under vacuum or by spraying with fine mist of water. Direct immersion of dry soil in water at atmospheric pressure causes disintegration of large aggregates into smaller ones, causing error in the estimation of water stable aggregates.
- The level of water in the tank should be maintained such that bottom of the top (5mm) sieve is immersed in the water during up stroke.

Observations and Calculations :

weight of fresh sample, g = W

weight of oven-dried sample, g = y

weight of aggregates collected from each sieve, g = w_1

Sieve dia (mm)	Mean diameter of aggregates (d_i) mm	Weight of aggregates (w_i) g	Mass of sand (w_q) g	Corrected mass of aggregates $(w = w_i - w_q)$ g

A graph plotted for aggregates weight against aggregate size gives the size distribution of aggregates. To represent the aggregation status of soil by a single value, the following indices can be calculated.

(a) Mean weight diameter $= \sum_{i=1}^{n} d_i \, w_i$

Where *n is* number of size fractions (the finest fraction that passes through the finest sieve inclusive), d_i is the mean diameter of each size range, w_i is the weight of aggregates in that size range as a fraction of the total dry weight of the sample (W) taken.

(b) % Aggregates > 0.25 mm dia $= \dfrac{\text{Weight of aggregates} > 0.25\text{mm dia}}{\text{Weight of the soil}} \times 100$

(c) % Aggregates stability

$$\% \text{ Aggregates stability} = \dfrac{\text{Weight of aggregates} > 0.25 \text{ mm dia} - \text{wt. of sand}}{\text{Weight of the sample} - \text{wt. of sand}} \times 100$$

Result :

M.W.D is ....

% Aggregates > 0.25 mm dia ....

% Aggregates stability ....

Exercise - 2.6

Penetrability

Soil penetrability is a measure of the ease with which an object can be pushed or driven into the soil. The ease with which the body is driven into soil depends on the opposing force or resistance offered by the soil. Any device used to measure resistance to penetration is called penetrometer. Determination of penetration resistance or mechanical impedance offered by the soil is very much essential to know about the condition of the soil *i.e.,* whether the soil is a hard soil, loose soil, moist soil *etc*. It helps us to know about the root penetration into soil. Similarly seedling emergence is associated with the mechanical impedance of over burden soil. Penetration resistance is influenced by soil and probe characteristics. Penetrometer factors affecting penetration resistance are cone angle, diameter, roughness and rate of penetration. Soil factors influencing penetration resistance are matric potential (water content), bulk density, soil strength parameters and others.

There are two principal types of penetrometers and ways of measuring penetration resistance.

1. Dynamic (driving the penetrometer into soil with hammer or by keeping the weights)
2. Static (the probe is pushed steadily into soil without any impact).

There are different types of penetrometers : cone penetrometer, friction sleeve penetrometer and pocket penetrometer. Depending on the accuracy, necessity, condition of the soil any one of them will be used. Cone penetrometer is used to measure the penetration resistance of the soil at different depths where as pocket penetrometer is used to measure the resistance at soil surface.

Aim :

To measure the penetration resistance using cone penetrometer and pocket penetrometer.

Principle:

The force required to push an instrument in to the soil depends on the amount of resistance offered during the penetration which is directly proportional to the strength of the soil.

Measurements using Cone Penetrometer

Apparatus :

Cone penetrometer (Fig. 2.11) with calibration chart.

The cone penetrometer consists of a handle, proving ring and dial guage, 100cm rod graduated at 10cm intervals and stainless steel cone. The cone has a base area of 6.45 cm^2 or 3.2 cm^2 and has a tip angle of 30 to 60 degrees.

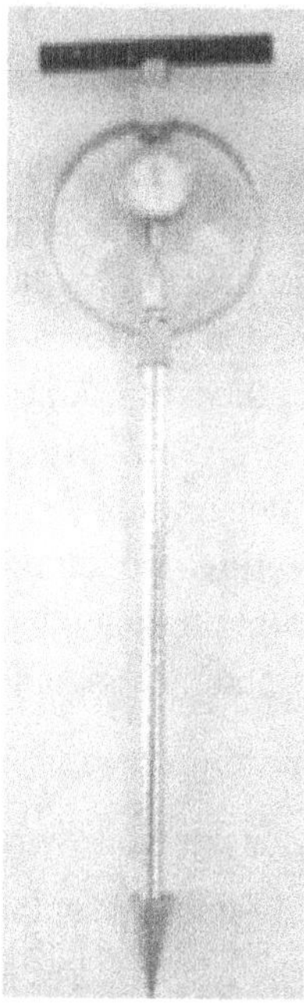

Fig. 2.11 Cone penetrometer.

Procedure :

    (a) Preparation of calibration chart

Set the gauge of the penetrometer at zero reading. Hold the penetrometer vertically on a smooth leveled hard surface and add weights on the top of the handle in the increasing order and record the gauge readings. Draw a graph between the weights (on X-axis) and gauge reading (on Y-axis). Graph will be a straight line passing through origin.

    (b) Determination of resistance offered by soil for penetration

The procedure is best performed by two persons. An operator who presses the cone penetrometer into the soil and calls out times for readings and a recorder to record the readings.

- Select the test location and prepare a flat, clean soil surface for the determination.
- Set the dial gauge to zero position.

- Hold the penetrometer in a vertical position and push the cone point slowly downward into the soil at a uniform rate (it should take 15 sec to reach a depth of 60cm).
- Take readings of the dial gauge at desired vertical increments of 5 or 10 cm, marked on the rod or the penetrometer.
- Repeat the determination several times at each location to obtain at least three sets of consistent and reliable readings.
- Average the dial gauge readings obtained at each depth for atleast 3 tests.
- Using the calibration chart convert the average dial readings to penetrometer force in pounds or kilograms.

Precautions :

- Space the individual penetrations so that they do not interfere with one another but should not be too far apart because variation in reading could occur due to spatial variation in the soil strength.

Observations and Calculations

S.No.	Soil depth (cm)	Dial guage reading

Result :

Penetration resistance is.........

Meaurements using Pocket Penetrometer

The pocket penetrometer is a hand-operated, calibrated-spring penetrometer. The deformation of the spring, as the piston needle is pushed into soil in the prescribed manner, has been correlated with strength of soil in kg/cm^2. The values are calibrated directly on a scale on the piston barrel. It is commonly used to evaluate crust strength at the soil surface in agricultural fields.

Apparatus :

Pocket penetrometer (Fig. 2.12).

Direct-reading pocket penetrometer has a diameter of 20 mm and a piston needle diameter of 6.35 mm (0.25 inch).

Fig. 2.12 Pocket penetrometer.

Procedure and Calculation

- Grip the handle, and push the piston needle, with steady pressure, vertically into the soil surface until penetration reaches the calibration groove (approximately 6.25 mm).
- The scale has a sliding indicator which holds the reading when the piston is released. Read the compressive strength, in kg/cm^2 on the penetrometer scale.
- Clean the needle, and return the sliding indicator to its zero position.
- Repeat the test several times in different areas to find an average value for compressive strength.

Results :

Penetration resistance of the surface soil is………

Date : ______________

Exercise - 2.7

Soil Color

Soil color is one of the most important characteristics of soil which is frequently used to describe a soil than any other property. Soil color has no direct effect on plant growth but an indirect effect on temperature and moisture. Color can be an indicator of the climatic condition under which a soil was developed or of its parent material. The productivity of soil can be judged from its color. Practically all colors occur in soils– white, red, brown, gray, yellow, and black. Even bluish and greenish tinges occur. Generally soil colors are not pure. Blue and green are not known to exist. Frequently two or three colours occur in patches. This is called "mottling". The color of the soil is a composite of the colors of its components. The effect of these components on the color of the soil is proportional to the total surface which is equal to their specific surface times their volume percentage in the soil. That means colloidal material plays an important role in the color.

Aim :

To measure the color of the soil using munsell color chart.

Apparatus :

Munsell color chart - It contains an orderly arrangement of 175 color combinations with notation for each color unit. The arrangement of the colors is according to the three basic components of color - hue, value, chroma. Hue refers to the dominant spectral color such as Red, Yellow, Blue *etc*. Value refers to the relative brightness or darkness of color with in a scale of (0-10) as compared to absolute white. It refers to gradation from white to black. Chroma refers to the relative purity of a color with in a scale ranging from (0-20). Degree of departure from neutral grays or white. Chips of the main colors occurring in soils are placed on card board charts. Each of these chips is designated by a numerical system that includes hue, value and chroma. A circular hole between two adjoining and similar color chips permits one to put a soil ped next to these chips and compare the colors. After the best match is found the soil color is described by the number of the chip to which it corresponds. For *eg:* In 2.5YR 5/6 : 2.5YR represents hue; 5 represents value and 6 represents chroma.

Principle :

Color is the visual impression created by the object which is carried by electromagnetic radiations. The wave length of visible light extends from 0.3 to 0.75 microns. The effect of the light of the various wave lengths affects the eye very differently. These different impressions are called colors. The color of an object depends upon the kind of light which it reflects to the eye. The impressions received by the human eye or the color of the object depends upon variations in the impressions had by different people, the proportion of the object and quality of light that it reflects and the texture of the object (finer the particles lighter is the color).

Procedure :

- Take the soil sample preferably in small aggregates.
- Place the dry aggregate under the holes provided between the colored chips to match the colour in the munsell chart.
- Note down the respective hue given at the right hand corner of the chart, value given on the left hand corner of the chart and chroma given at the bottom of the chart of the soil colour in dry state.
- Then moisten the aggregate with few drops of water.
- Place the moist aggregate under the holes provided to match the colour in the munsell chart.
- Note down the respective hue, value and chroma values of the soil aggregate in moist state.

Precautions :

- To avoid confusion use white mask for dark colored soil and dark for other soils while confirming the colour.
- Consider natural surface of weakness of the aggregate for identifying the colour.

Result :

Colour of the soil is ………..

Chapter 3

Soil Hydraulic Properties

Date : _______________

Exercise - 3.1

Soil Water Content by Gravimetric Method

Knowledge of soil water content is required to assess the extent of availability of water to plants, to work out optimum soil water regimes, depletion patterns and consumptive use for various crops, to schedule irrigation and calculate depth of irrigation water needed to bring the soil to a specified water content, to study soil water movement and relate certain physiological processes of plants with soil water content. Methods followed generally, for the estimation of soil moisture content depending on the accuracy, need and nature of data required, are gravimetric method, electrical resistance method, neutron probe method and g-ray attenuation method.

Aim :

To estimate the moisture content present in the soil by gravimetric method.

Principle :

A wet soil is dried in oven and the loss of water on drying is determined. This requires weight of the wet soil and weight of the dry soil. Drying the soil is done in an oven at 105^0C temperature of to a constant weight.

Apparatus :

Soil auger, aluminum boxes, oven and balance

Procedure :

- Collect the moist soil samples with the help of an auger from the field.
- Transfer the moist soil sample to a pre-weighed aluminum box.
- Weigh the sample along with the box after taking them to laboratory.
- Place the box with lid off in an oven at a temperature of 105 ^{o}C to dry the soil to a constant weight.
- Remove sample from the oven.
- Replace the lid and place the box in a desiccator.
- Upon cooling to the room temperature determine the weight of dry soil and the moisture box.

Precautions :

- Take weight of the moist soil sample collected without loosing much time.
- Place the moisture boxes with their lids removed in the oven for drying the soil.

Observations and Calculations :

weight of moisture box + moist soil, $g = w_{bsm}$

weight of moisture box + dry soil, $g = w_{bsd}$

weight of empty moisture box, $g = w_b$

weight of water in the soil, $g = w_{bsm} - w_{bsd}$

weight of dry soil, $g = w_{bsd} - w_b$

$$\text{Per cent water (dry weight basis)} = \frac{w_{bsm} - w_{bsd}}{w_{bsd} - w_b} \times 100$$

Result :

Moisture content of the soil sample on dry weight basis is ____________

Excerise – 3.2

Soil Water Content Using Neutron Probe

Aim : To estimate moisture content in the soil using neutron probe

Principle : The neutron method is an indirect non-destructive procedure of finding soil moisture by volume. When fast neutrons (kinetic energy of about 2 MeV) are emitted from a source like radium or beryllium, they interact with the surrounding matter by elastic and non-elastic collisions and lose their energy down to that of thermal neutrons (about 0.025 eV). Because small atoms are the best moderators of fast neutrons, the most efficient atom for neutron moderation is hydrogen. Owing to the two hydrogen atoms in its molecule, water is an excellent moderator of fast neutrons. Higher is the water content in the soil, greater is the number of slowed down neutrons. The low energy neutrons are counted by a meter. The number of neutrons counted is proportional to the hydrogen nuclei and hence the volumeteric water content in the soil.

Appratus : Neutron moisture meter, aluminum access tubes, rubber stoppers, access tube caps, soil auger, film badges, aluminum boxes, oven, balance.

Procedure

A	Calibration Procedure

- The instrument is to be calibrated at different moisture contents ranging from dry to wet. This can be done in the dry season by providing different amounts of water to different plots.
- Depending on the type of the probe used, collect soil core samples of 10 cm long using suitable soil tube augar starting fom 10 cm soil depth to a maximum rooting depth.
- Soil samples are to be transferrred to the aluminum cans and determine the water content by gravimetric method.
- Insert the aluminum access tube, one end of which is plugged with a rubber stopper into the soil hole such that it fits snugly to the maximum depth leaving only 25 cm of the tube above the ground level.
- Place the probe unit over the access tube preparatory to lowering it into the hole.
- Take four or five standard counts while the source is still in the shield for a fixed time.
- Now slowely lower the source to the middle of each layer sampled, *i.e.*, at 10, 20, 30, 40, 50 cm etc soil depths and take the actual probe counts for fixed time.
- Repeat the process of soil sample collection, installation of access tubes, and probe readings in the plots of different moisture contents.

- Repeat this process several times to get sufficient data points for developing a calibration curves.
- Determine the bulk density of each layer from each plot by taking core samples (as in exercise 2.3.)
- Multiply the gravimetric moisture content by the bulk density of each horizon to calculate the volumetric moisture content.
- Calculate the count ratio (CR) by dividing actual counts from each soil depth by the mean standard count.
- Develop a calibration equation for the soil by regressing volumetric moisture content on the count ratios as a dependent variable. The calibration equation thus developed is used for estimating water content of the soil if count ratios are known.

**B	Taking Probe readings and estimating soil moisture**

- Install neutron probe access tubes at the required places following the above procedure. The number of access tubes per each plot is determined by the size of the plot and soil variability.
- Clean the access tubes. Check the tubes with a dummy probe so that it moves freely in the tube.
- Place the neutron probe over the tube and take four or five standard counts while the probe is still in the shield.
- Lower the source into the tube and take readings at every 10-cm depth intervals, starting at a soil depth of 10 cm.
- Calculate count ratio (CR) by dividing the actual counts by the mean standard count.
- Calculate volumetric moisture content by substituting the values of CR in the calibration equation.

Precautions :

- Neutron probe is not suitable to determine the moisture content of 0-5 cm soil layer. So determine the moisture content by the gravimetric method and multiply with bulk density to obtain the volumetric water content.
- Separate calibration curves may be needed for different soil layers of the same soil profile if they markedly differ in soil properties.
- Follow the safety rules supplied by the manufacturer, along with the instrument. But important precautions are the following:
  - Keep the probe in the shield at all times except when it is lowered into the soil for measurements.
  - Personnel who operate the probe should reduce exposure to the small radiation escaping from the shield by maintaining a distance of a few metres between them and the probe, except when changing its position.
  - Transport the probe in the back of a truck, or in a car trunk.
  - Operators should wear a film badge at waist level.
  - When the probe is not in use, lock it in a storage room away from people.
  - Probe maintenance should be performed by personnel trained in the use of radioactive equipment.

Exercise - 3.3

Soil Water Potential

Information on soil water content is of primary interest for many studies. However, for studies involving water transport and storage in soils and soil-water-plant relationship the energy status of soil water is required. Measurement of soil moisture tension/potential, field capacity and drainage under actual field conditions are important properties. These are measured by tensiometry system. Measurement of soil moisture tension has practical applications in irrigation scheduling, root zone delineation, hydraulic gradient measurements *i.e.*, drainage. In coarse textured soils where 80 to 90% of moisture is within the tensiometer range such measurements will be of great help in field condition.

Aim :

Measurement of soil water potential using tensiometer

Principle :

Tensiometer is used to obtain energy status of water in soil directly and water content indirectly assuming that the former is in equilibrium with soil water. As the water content of the soil surrounding the water-filled porous cup decreases, the energy level of soil water decreases relative to that of the water in the tensiometer cup and water moves out of the tensiometer through the pores of the tensiometer cup into the soil. Section is created in the tensiometer. The tension or potential reading in the tensiometer guage is equal to soil water tension. Tensiometer can be used to measure the tension upto 0.85 bars.

Apparatus :

Tensiometer, soil auger

A tensiometer (Fig. 3.1) has the following components :

- Pressure measuring device (vaccum or suction gauge).
- Porous cup (95 mm long, 29 mm outer diameter and 1.6 mm wall thickness).
- Acrylic tube 13 mm diameter.
- Deaerated water (boiled and cooled distilled water).
- Rubber caps for the tube.

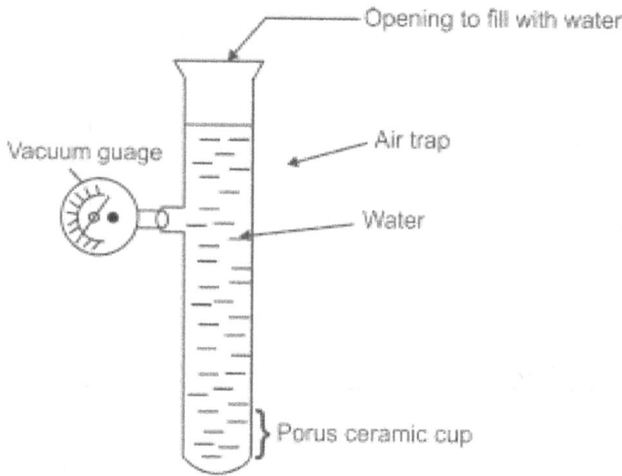

Fig. 3.1 Tensiometer.

Procedure :

A. Testing of tensiometer

For proper working, the tensiometer assembly should be air tight. For this purpose each tensiometer should be tested prior to field installation. Place a tensiometer in a bucket of water after takingout the lid. Water enters into the tensiometer through the porous cup.   If water does not move into the cup (showing low conductance), it means that walls of the cup are choked. For removing this defect, treat the cup with warm 1 : 4 HCl solution.   If such treatment is unsuccessful, replace the cup with a new one and test it. Response time should also be tested by developing a suction of 0.6 to 0.8 bars in the instrument by evaporation from the cup and then submerging the cup in water. If there is any leakage in the system or the cup is damaged suction will not develop in the tensiometer.

B. Installation

- For installing tensiometer, make a hole to the desired depth with the help of a tube auger or a screw auger of the same diameter as that of tensiometer barrel.
- Insert the tested tensiometer into the soil to the desired depth gently pushing and twisting. If there is a gap around the tensiometer fill the gap with the soil slurry and see that there is no gap around the tensiometer.
- After installation, fill the tensiometer with deaerated distilled water. Fit the cap on the tensiometer pipe firmly after deaerating the whole system. Leave the tensiometer for sufficient time in the soil till the hydraulic equilibrium is reached (nearly 24 hours).
- Measure the reading in the vaccum guage.
- From the soil water characteristic curve (calibration curve) soil water content with respect to the soil water potential–suction can be derived.

Precautions :

- Porous cup should be saturated with water before installing the tensiometer.
- All joints in the tensiometer assembly should be air tight.

Observations and Calculations :

Guage reading =

Result :

Soil water suction =…………….

Soil water content from the calibration curve with respect to the measured soil

water suction = …………….

Exercise - 3.4

Soil Moisture Constants

Field capacity (FC) is the amount of water held in the soil after excess gravitational water has drained away and after the rate of downward movement has materially decreased, which usually takes place in 1-3 days after a heavy rain or irrigation. At field capacity air occupies macro pores and the micro pores are filled with water. Moisture content at field capacity is considered as the maximum limit of soil moisture available to the plants. The soil moisture at field capacity is around 0.1 bars tension in course textured soil and 0.33 bars in fine textured soil, field capacity is influenced by other soil propels like texture, structure etc.

Permanent wilting point (PWP) is defined as the water content of a soil when indicator plants growing in that soil wilt and fail to recover when placed in a humid chamber. The soil water suction at this permanent wilting point is around 15 bar. Permanent wilting point is generally considered as the lower limit of available water to the plants. However permanent wilting point is influenced by plant and atmospheric factors. Among the soil factors, texture influences permanent wilting point to a grater extent. Fine textured soils have higher permanent wilting point than coarse textured soils.

Under field conditions field capacity and permanent wilting point are not the constant values of soil. Nevertheless these values are used even now for practical purpose of irrigation management. Hence determination of these values is very important.

The **available water capacity** (AWC) of a soil is the amount of water retained in the soil that can be removed by the plants. For field soils, the AWC is estimated by the difference in soil water content between FC and PWP. Though considerable moisture is present below the PWP, particularly in fine textured soils, it is held so tightly in the soil that plant roots cannot extract it rapid enough to meet the transpiration demand and so the plant wilts. Similarly, the water above the FC is easily drained out due to gravity, within a short period of time and hence considered as not much available to the plants"

$$AWC = FC - PWP$$

Several methods are available to determine the water content of soil at FC and PWP of soil, of which pressure plate apparatus method is an important laboratory method.

Aim :

Determination of field capacity, permanent wilting point and available water using pressure plate Apparatus.

Principle :

After the porous ceramic plate of the pressure plate apparatus and the soil sample are completely saturated with water, it is mounted in the pressure vessel. Air pressure is released into the vessels. As soon as the air pressure inside the chamber is raised above atmospheric pressure, the higher pressure inside the chamber forces excess water through the pores in the ceramic plate and out through the outlet stem via the passage afforded by the screen. Equilibrium is reached in terms of soil water suction and the applied air pressure. Out flow of water from the samples stop at equilibrium when the air pressure in the extractor is increased, the flow of soil moisture from the samples starts again and continues until a new equilibrium is reached. When the out flow of water stops at a particular applied pressure, the soil sample are taken out and the moisture content is determined.

Apparatus :

Pressure plate apparatus (Fig. 3.2), metal or plastic or rubber rings 1 cm height and 5 cm diameter (sample retainer rings), balance, oven, moisture boxes, soil sampling auger.

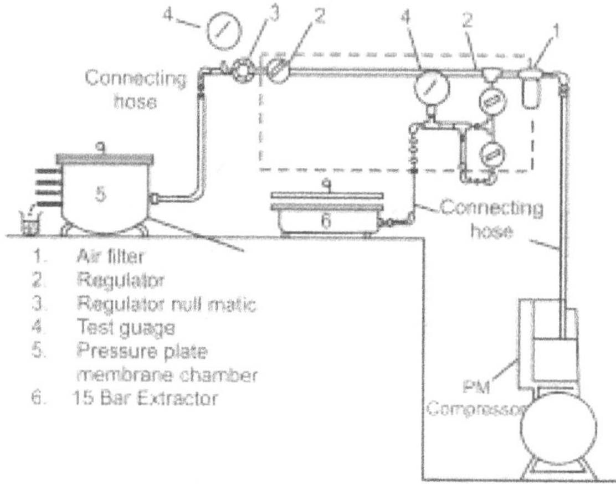

Fig. 3.2 Pressure plate apparatus.

Procedure :

- Take undisturbed soil samples from the field with sample retainer rings.
- Put the soil samples along with rings in duplicate on the ceramic plate (1 bar plate for FC and 15 bar plate for the PWP should be used)
- While placing the samples on ceramic plates, make sure that soil cores have good contact with the plate. This facilitates wetting of the soil sample from below.
- Saturate the samples by placing the plate with the rings in a trough of water for at least 24 hours. Water should be just enough to reach the upper edge of the rings.
- After saturation remove the excess water from the ceramic plate with pipette, transfer the plate into the pressure chamber and place it on the supporting system.

- Connect the nylon tube and rubber sleeve to the outlet pipe of the pressure plate apparatus.
- Close all unused outlets with the provided plug bolts and make sure that 'O" ring is in place..
- Fix lid on the pressure chamber with provided nuts and bolts.
- Apply pressure slowly. As the pressure builds up inside the chamber, water will come out from outflow tubes
- When soil water pressures equilibrate with air pressure, flow of water ceases. Equilibrium is usually reached in about 3 days for low pressures and 5 days for high pressures depending on texture and height of sample.
- After equilibrium, release the pressure in the chamber by shutting off the regulator gently.
- When the extraction is completed, then close the drainage pipe, and remove the samples.
- Remove clamping bolts and lid to open the chamber.
- Transfer samples to moisture boxes. Take wet weights and dry them in oven at 105^0C for 24 hours and determine the water content.
- This is the water content at the pressure maintained.
- For field capacity the pressure in the chamber is maintained at 0.1 bar for coarse textured soil and at 0.33 bar for fine textured soils. For PWP the pressure is maintained at 15 bars.

Soil moisture characteristic curve :

Soil moisture contents at different values of pressure like 0.1, 0.3, 1.0,5.0,10 and 15 bars can be determined. The curve showing relationship between moisture content and suction is known as soil water characteristic curve. It is important to know the potential storage of water in soils and its release characteristics to have an idea of water availability to plant roots as water stress develops in plants. As soils differ in texture and structure, they have different water retention and release characteristics. These can be determined in the laboratory using pressure plate apparatus.

Precautions :

- Make sure that the pressure chamber measurements are done in a room with minimum temperature fluctuations.
- Select the pressure chamber system and the porous plates depending upon the range of pressures to be applied.
- If repacked cores are used, the bulk density must be chosen to match the insitu bulk density.
- Prevent air leakage from pressure plate assembly.
- Increase pressure in the chamber gradually to the desired level
- Much time should not be lost while taking wet weight of the sample after removing from the pressure chamber.
- Pressure should be released slowly at equilibrium point after closing the out flow tube.

Observations and Calculations :

At 0.1/0.33 bar

Empty box, g = w_1

Weight of wet sample + box, g = w_2

Weight of dry sample + box, g = w_3

Water content on dry weight basis, g/g =

At 15 bar

Empty box, g = w_1

Weight of wet sample + box, g = w_2

Weight of dry sample + box, g = w_3

Water content on dry weight basis, g/g = $\dfrac{w_2 - w_3}{w_3 - w_1}$

Available water capacity (AWC_w) = water content between FC and PWP

= [water content at 0.1or 0.3 bar] - [water content at 15 bar]

Profile water storage capacity = $S^n_{i=1}$ [(FC – PWP) × bulk density × thickness of the layer]

Calculation of profile available water capacity

Depth increment (cm) (1)	AWC_w (g / g) (2)	Bulk density (g / cm^3) (3)	AWC_v (cm^3 / cm^3) (4) = (2) × (3)	Layer thickness (cm) (5)	Depth of water (cm) (6) = (4) × (5)
Total					

Result :

Water content at field capacity (0.1/0.33 bars) = ...g/g

Water content at wilting point (15 bars) = ...g/g

Available water capacity = ... g/g

Profile water storage capacity = ... cm/m

Exercise - 3.5

Water Intake Rate

The downward entry of water into the soil through the soil surface is called infiltration. It is an important soil property because it partitions rain into soil water and runoff. It depends on many factors such as soil texture, antecedent moisture content, surface cover, and soil management. Infiltration characteristics of soil are of practical significance in irrigation, soil and water conservation and watershed management.

Aim :

To measure the water intake rate of the soil using double ring infiltrometer.

Principle :

The main principle is to measure the amount of water entering the soil profile as a function of time. During infiltration, appreciable lateral movement of water may also occur. To avoid error due to this lateral movement, two concentric iron rings (infiltrometers) are used. Water level in both the rings should be kept nearly equal. The rate of fall of water level in the inner ring is measured.

Apparatus :

Double ring infiltrometer (Fig. 3.3) - two rings made from 14-16 gauge iron sheet rolled into a cylinder, Outer ring of 60 cm and inner ring of 30 cm diameter, both 30cm height; spade; buckets; polythene sheet; watch; hook gauge/scale; driving plate; hammer.

The lower edges of the rings are sharpened to facilitate easy drive of the rings. The top of the rings are provided with sturdy rims.

Procedure :

- Describe the texture, surface condition, structure, compaction, salt content and layering sequence in the profile.

Fig. 3.3 Infiltrometer rings.

- Determine the initial soil moisture content.
- Install infiltrometer rings in a uniform and nearly level plot to a depth of 15cm.
- Pond 10-15 cm of water in outer as well as inner ring.
- Record the fall of water level in the inner ring with hook gauge at 1,3,5,10,20,30,40,60,80,100,120 minutes and thereafter on hourly basis till the water intake is constant. However, the time intervals of observation can be varied according to objectives of study and soil permeability. More water should be added into the rings when water level falls by 4 to 5 cm in order to check drastic water level fluctuations which may affect constant intake rate otherwise.
- Using scale or hook gauge record the water level and time just before and after reponding. Keep the intervals between these two observations as short as possible to avoid errors caused by intake during the refilling period.
- Plot the infiltration rate and cumulative infiltration as function of time.

Precautions :

- Drive the rings straight down with minimum soil surface disturbance.
- Cover the soil surface in the rings with polythene sheet while pouring water in the rings for the first time to avoid disperssion surface soil and clogging of pores.
- Try to keep level identical in the inner as well as outer ring.

Observations and Calculations :

Location

Texture of the surface

soil surface conditions (kind and extent of vegetation, crust, salts)

Layering sequence in the profile

Antecedent soil water content

Specific problem, if any (erosion, salinity, ponding)

Area of the inner ring

Result :

Infiltration rate is …………mm / hr

Inference :

Soil comes under the category ……………… (as per the table below)

Tabulate the observations and calculations in the following format

S.No.	Cumulative Time (t) (min)	Water level reading, r (cm)	Cumulative intake (cm)	Infilltration rate, dr/dt, (cm/min)
1	0			
2	1			
3	3			
4	5			
5	10			
6	20			
7	40			
8	80			
9	120			
10	300			

Classification :

Descriptive term	Infiltration rate	
	in./hr	mm/hr
Very rapid	>10.00	>254
Rapid	5.00 - 10.00	127 - 254
Moderately rapid	2.50 - 5.00	63 - 127
Moderate	0.80 - 2.50	20 - 63
Moderately slow	0.20 - 0.80	5 - 20
Slow	0.05 - 0.20	1 - 5
Very slow	<0.05	<1

Exercise - 3.6

Saturated Hydraulic Conductivity

Hydraulic conductivity of a soil may be defined as the ability or capacity of the soil to transmit or transfer the water from higher pressure points to lower pressure points. Saturated hydraulic conductivity pertains to the conductivity of soil when all pores, macro as well as micro, are filled with water and unsaturated hydraulic conductivity refers to the conductivity when pores are partially filled. The saturated hydraulic conductivity of soil refers to the readiness with which a saturated soil transmits water through its body and is expressed as length per unit time. It depends on the properties of both soil and water. The soil factors affecting the hydraulic conductivity are pore geometry, mineralogical composition of the soil, stratification, presence of entrapped air and the microbial activity in the soil; and water properties effecting hydraulic conductivity are its viscosity and density. Therefore saturated hydraulic conductivity is to be determined collecting undisturbed soils in the permeameter. Hydraulic conductivity of various textured soils vary from 0.001cm/hr (fine textured) to 25 cm/hr (coarse textured).

Aim :

To determine saturated hydraulic conductivity of soil using constant head method.

Principle :

It is based an Darcy's law which states that if a constant water head is maintained on one end of a saturated column of soil of length (l), the volume of water (V) passing through the other end per unit cross- sectional area (A) of the soil column per unit time (t) will be directly proportional to the hydraulic gradient (H/l) across the length of the soil column. Thus :

$$V = K \, (H/l) \, A \, t$$

According to Darcy's law the proportionality constant, K in the above equation is the hydraulic conductivity of the soil. The symbol H stands for the difference in total head between inflow and outflow ends of a column.

Apparatus :

Metal cores (permeameters) of about 5 cm inside diameter and 15 cm length, a wooden or iron stand with funnels embedded for supporting the cores, a water reservoir with siphon tubes arrangement for maintaining a constant water head on the soil surface in the cylinder (Fig. 3.4), a stop watch, graduated cylinders, filter paper discs, muslin cloth, rubber bands, scale, 500ml glass beakers.

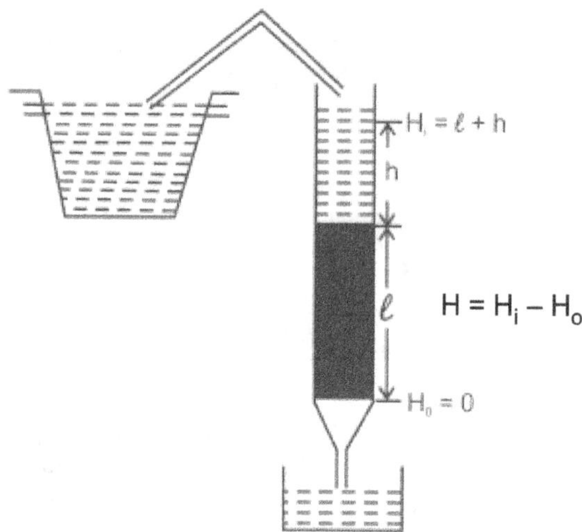

Fig. 3.4 Saturate hydraulic conductivity measurement device.

Procedure :

- Collect undisturbed soil core of about 5cm length in the permeameter or cylindrical metal tube. Cut the excess soil with a sharp knife. Keep a filter paper at the bottom and then close with porous cloth with the help of rubber band.
- Measure the total length of the permeameter with scale.
- Keep it for saturation in a tray of water.
- Measure the gap above the soil.
- Place the core on the stand and start siphoning the water to maintain a constant head of 2-3 cm of water on the top of the soil by siphon tubes and constant water level arrangement as shown in the Fig. 3.3.
- When a steady flow is obtained, start collecting the percolate in a graduated cylinder or a beaker.
- Measure the volume of percolate collected in a known time.
- Record atleast three consecutive constant readings till the flux is constant.
- Measure the exact water head on the soil surface with the help of a scale and then discontinue the experiment.
- Measure the length of water column above the soil.
- Measure the inner diameter of the cylinder.

Precautions :

- Take care to maintain constant head.
- Avoid overflow of water.

Observations and Calculations :

Inner diameter of the core (permeameter) cm = d

Inner radius of the core cm = r

Cross-sectional area of the permeameter.cm^2 = A = p r^2

Depth of water above the soil .cm = h

Length of soil column. cm = l

Time for which percolate collected. min = t

Volume of percolate collected. cm^3 = V

Total head at the inflow, cm H_i = h + l

Total head at the outflow, cm H_0 = O

Hydraulic head difference, cm H = H_i - H_0 = h + l

Hydraulic gradient = (h + l) /l

Hydraulic conductivity, K in cm min^{-1} = (V/t) * l/HA cm min^{-1}

S.No.	Time of collection (min)	Volume of collected water (ml)
1		
2		
3		
4		

Result :

Saturated hydraulic conductivity of the soil is.........

Inference :

The soil falls under....................permeability class. (as the table below)

Permeability class and hydraulic conductivity of saturated soils

Permeability class	cm/h
Very slow	<0.108
Slow	0.11-0.54
Moderately slow	0.54-2.16
Moderate	2.16-6.12
Moderately rapid	6.12-12.6
Rapid	12.6-25.2
Very rapid	>25.2

Chapter 4

Thermal Properties

Date : _______________

Exercise - 4.1

Soil Temperature

Soil temperature refers to the intensity of heat in the soil body expressed as degrees. It is an important edaphic factor that controls physical, chemical and biological processes occurring in soil and influences plant growth. Soil temperature depends upon type, structure, water content, surface cover, topography, intensity of incoming solar radiation, absorbance capacity, variations in thermal capacity, biological activities, concentration of salts and evapotranspiration *etc.*   Different types of thermometers like mercury or liquid in glass, bimetallic, bourdon and electrical resistance thermometers, thermocouple and thermistor are available one of which can be chosen depending upon the nature of study and the accuracy desired.

Principle :

Soil temperature can be measured by considiring by its influence on some properties of matter that respond to variation in intensity of heat in body of the matter. Changes in properties of matter given in Table 4.1 have been found useful for the construction of thermometers for the temperature measurements.

Table 4.1 Thermal properties of different substances used for measuring temperature

Thermo-metric substance	Property
Mercury or liquid in a glass capillary tube	Change in Volume
Bimetallic strips	Change in Length
Platinum or other wires	Change in Electrical resistance
Thermocouple	Change in Thermo EMF
Gas or vapour at constant volume	Change in Pressure

Liquid in glass capillary tube thermometer

Generally Mercury is used in this thermometer. The response of Hg to the temperature in terms of change in the length of the mercury column in the glass capillary tube has been used in the construction of this thermometer. Glass capillary tube having a bulb at its bottom is enclosed in a glass tube which is protected by metal casing with a wooden handle at its top. Scale is calibrated just like in ordinary mercury in glass tube thermometer. The metal casing is provided with sharpened metallic narrow tube at its bottom with some perforations which facilitate easy drive and good thermal contact of the sensitive element of the thermometer.

Bimetallic thermometers

The name of this thermometer indicates the presence of two different metals having different coefficients of linear expansions. The two metals are taken in the form of strips and they are fused together. When this combination of metals is exposed to the heat energy, due to the difference in the coefficients of linear expansion, the response of the combination of metals will be reflected as some deformity in its shape. This deformity which is linearily related to temperature will be magnified and conveyed to the measuring device whose scale is calibrated directly in degrees centigrade.

Electrical resistance thermometers

Every conductor offers some resistance to the flow of electricity. The resistance offered not only depends on the nature of the conductor but also depends on temperature. Except for carbon, the resistance offered by all conductors increases with increase of temperature. Even though platinum is costly element due to its reliability, it is generally used to construct electrical resistance thermometers. The element like platinum enclosed in glass will be driven up to the point of measurement and will be connected as fourth arm of the Wheatstone's bridge (which is meant for the measurement of resistance). The resistance thus measured will be correlated with the standard linear relationship between resistance and temperature.

Thermocouple

Thermocouple is an arrangement of two different metals taken in the form of wires and fused at their ends to form a net work. When the two junctions are maintained at two different temperatures due to temperature difference some thermo EMF will be developed and thermo electricity will be generated. Thermo EMF produced depends on temperature difference between two junctions which can be studied with the help of calibration curve. When it is used as soil thermometer practically one junction will be kept in the melting ice so that its temperature remains always at zero. The other junction is burried in the soil at the depth under consideration. Due to the temperature difference thermo EMF will be developed which can be measured with the help of potentiometer and it will be taken on Y-axis of the calibration curve and the temperature of the soil can be measured from the intercept on X-axis.

Aim :

To measure the soil temperature at different depths of soil.

Apparatus :

Mercury in glass thermometer, bimetallic thermometer (Fig. 4.1).

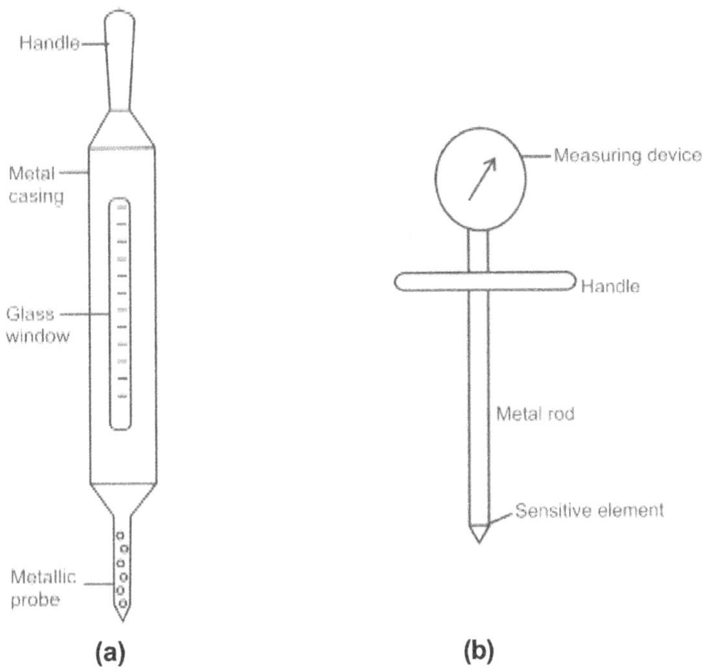

Fig. 4.1 Different Thermometers : (a) Liquid (Hg) type (b) Bimetallic

Procedure :

Drive the thermometer into the soil upto the desired depth and ensure that the sensitive element of the thermometer has good thermal contact with the soil. Usually liquid in glass tube thermometers enclosed in steel tube will be released in the holes provided by either crow bar or soil auger upto the depth and measurements will be made. Thermometers are to be left in the soil for 24 hours till thermal equilibrium is established. To study temperature variations with time, readings are to be taken at regular intervals of time.

Precautions :

- A thermometer must be properly calibrated and installed for reliable results.
- Thermometer when installed must be in good contact with soil and shielded from direct solar radiation.
- Readings should be taken at the same time to avoid errors because of diurnal variations.

Observations :

With mercury bulb thermometer

S.No.	Depth of the point	Temperature recorded

With bimetallic thermometer

S.No.	Depth of the point	Temperature recorded

Result :

- The temperature measured with mercury bulb thermometer at 5 cm depth and 10 cm depth are __________ and __________, respectively.

- The temperature measured with bimetallic thermometer at 5cm, 10cm and 15cm depth are ________, ________ and ________, respectively.